Bugs

Walking Sticks

by Trudy Becker

FOCUS READERS

PIONEER

www.focusreaders.com

Focus Readers is distributed by North Star Editions:
sales@northstareditions.com | 888-417-0195

Produced for Focus Readers by Red Line Editorial.

Photographs ©: Shutterstock Images, cover, 1, 8, 10, 12, 14, 17, 18, 20; iStockphoto, 4, 6

Library of Congress Cataloging-in-Publication Data
Names: Becker, Trudy, author.
Title: Walking sticks / by Trudy Becker.
Description: Lake Elmo, MN : Focus Readers, 2023. | Series: Bugs | Includes index. | Audience: Grades 2-3
Identifiers: LCCN 2022035857 (print) | LCCN 2022035858 (ebook) | ISBN 9781637394533 (hardcover) | ISBN 9781637394908 (paperback) | ISBN 9781637395639 (pdf) | ISBN 9781637395271 (ebook)
Subjects: LCSH: Stick insects--Juvenile literature.
Classification: LCC QL509.5 .B43 2023 (print) | LCC QL509.5 (ebook) | DDC 595.7/29--dc23/eng/20220818
LC record available at https://lccn.loc.gov/2022035857
LC ebook record available at https://lccn.loc.gov/2022035858

Printed in the United States of America
Mankato, MN
012023

About the Author

Trudy Becker lives in Minneapolis, Minnesota. She likes exploring new places and loves anything involving books.

Table of Contents

Chapter 1

Hiding in Plain Sight

A soft wind blows through the forest. A thin stick sways in the air. A hungry bird sees it but then flies away. The bird was fooled. The stick was actually an insect.

Walking sticks live in forests and grasslands. These insects are great at **camouflage**. Many **predators** think they are just sticks. That helps walking sticks stay safe.

Fun Fact

Walking sticks are also called stick insects.

Chapter 2

Walking Stick Bodies

Walking sticks might look like twigs. But they have many different body parts. Like all insects, they have six legs. They also have two **antennae**.

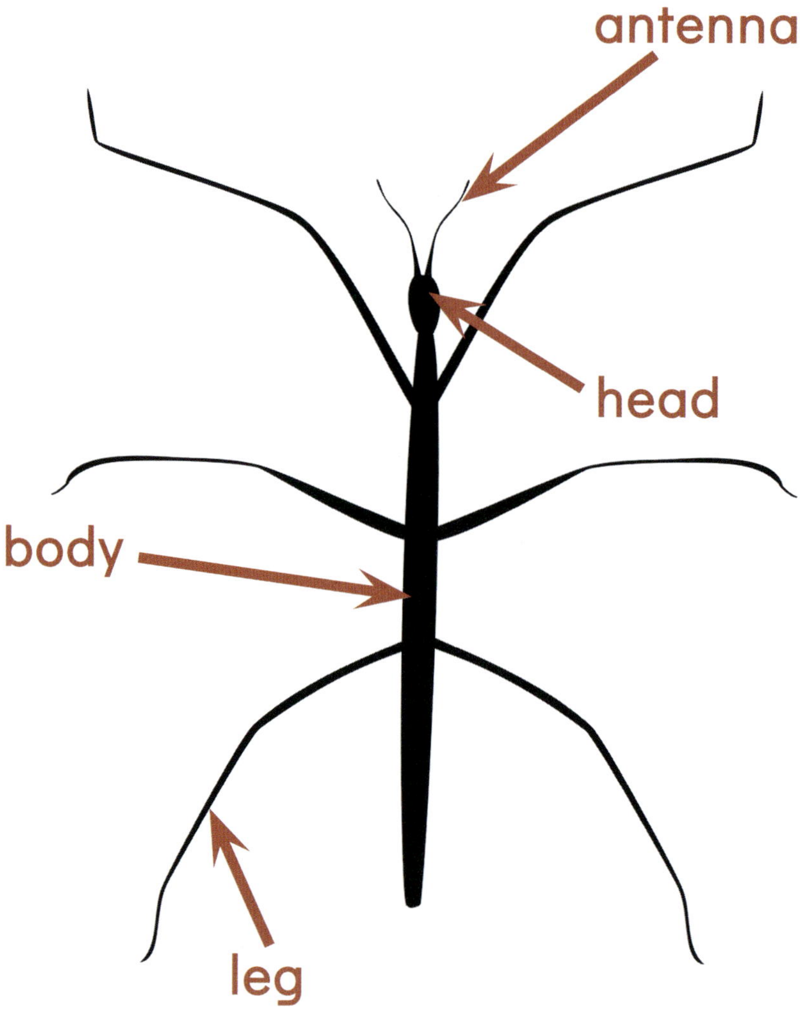
antenna
head
body
leg

Walking sticks have strong jaws, too. That's how they chew up their food. They have hard coverings to protect them. And some walking sticks even have wings.

Fun Fact

Walking sticks get new coverings as they grow. The old coverings fall off. Sometimes walking sticks eat the old coverings.

Chapter 3

Out of Danger

Walking sticks move in sneaky ways. They may walk slowly. They may sway back and forth. Predators think they are just sticks in the wind.

Walking sticks have other ways to avoid danger. Sometimes they play dead. Sometimes they drop their legs to escape. Some walking sticks can even let out bad smells to scare predators.

Fun Fact

Some walking sticks can be more than 20 inches (51 cm) long.

Males Not Needed

Most animals need females and males to have babies. But walking sticks don't. Female walking sticks can lay eggs on their own. Those eggs hatch into more female walking sticks. Sometimes males do **fertilize** eggs. Babies from those eggs can be male or female.

egg

Chapter 4

Walking Stick Life

Walking sticks mostly eat leaves. They are **herbivores**. But walking sticks are **prey** to spiders and birds. Bats eat them, too. Camouflage doesn't fool bats.

Walking sticks are good at hiding their eggs. Females put eggs in safe places, such as under leaves. Sometimes females drop each egg in a different place. That way, more **nymphs** can survive.

Fun Fact

Sometimes ants eat the coverings of walking stick eggs. Those eggs can stay safe inside anthills until they hatch.

FOCUS ON

Walking Sticks

Write your answers on a separate piece of paper.

1. Write a sentence that explains the main idea of Chapter 3.
2. Walking sticks are great at camouflage. How could you blend in with what's around you?
3. What body part do walking sticks use for eating?
    A. antennae
    B. jaws
    C. legs
4. Why does leaving eggs in different places give baby walking sticks a better chance to survive?
    A. Predators can only find one egg at a time.
    B. The eggs need lots of space to grow.
    C. One egg uses up an area's nutrients.

Answer key on page 24.

Glossary

antennae
Long, thin body parts on an insect's head. The parts are used for sensing.

camouflage
Colors that make an animal difficult to see in the area around it.

fertilize
To cause an egg or seed to start growing into a new young animal or plant.

herbivores
Animals that eat mostly plants.

nymphs
Baby insects.

predators
Animals that hunt other animals for food.

prey
Animals that are eaten by other animals.

To Learn More

BOOKS

Murray, Julie. *Stick Insects.* Minneapolis: Abdo Publishing, 2021.

Perish, Patrick. *Walkingsticks.* Minneapolis: Bellwether Media, 2019.

NOTE TO EDUCATORS

Visit **www.focusreaders.com** to find lesson plans, activities, links, and other resources related to this title.

Index

Answer Key: 1. Answers will vary; **2.** Answers will vary; **3.** B; **4.** A